Nanotechnology Science and Technology

ZnO Nano-Structures for Biosensing Applications: Molecular Dynamic Simulations

NANOTECHNOLOGY SCIENCE AND TECHNOLOGY

Additional books in this series can be found on Nova's website at:

https://www.novapublishers.com/catalog/index.php?cPath=23_29&seriesp=Nanotechnology+Science+and+Technology

Additional e-books in this series can be found on Nova's website at:

https://www.novapublishers.com/catalog/index.php?cPath=23_29&seriespe=Nanotechnology+Science+and+Technology

NANOTECHNOLOGY SCIENCE AND TECHNOLOGY

ZnO NANO-STRUCTURES FOR BIOSENSING APPLICATIONS: MOLECULAR DYNAMIC SIMULATIONS

SAFAA AL-HILLI
AND
MAGNUS WILLANDER

Nova Science Publishers, Inc.
New York

Copyright © 2010 by Nova Science Publishers, Inc.

All rights reserved. No part of this book may be reproduced, stored in a retrieval system or transmitted in any form or by any means: electronic, electrostatic, magnetic, tape, mechanical photocopying, recording or otherwise without the written permission of the Publisher.

For permission to use material from this book please contact us:
Telephone 631-231-7269; Fax 631-231-8175
Web Site: http://www.novapublishers.com

NOTICE TO THE READER

The Publisher has taken reasonable care in the preparation of this book, but makes no expressed or implied warranty of any kind and assumes no responsibility for any errors or omissions. No liability is assumed for incidental or consequential damages in connection with or arising out of information contained in this book. The Publisher shall not be liable for any special, consequential, or exemplary damages resulting, in whole or in part, from the readers' use of, or reliance upon, this material.

Independent verification should be sought for any data, advice or recommendations contained in this book. In addition, no responsibility is assumed by the publisher for any injury and/or damage to persons or property arising from any methods, products, instructions, ideas or otherwise contained in this publication.

This publication is designed to provide accurate and authoritative information with regard to the subject matter covered herein. It is sold with the clear understanding that the Publisher is not engaged in rendering legal or any other professional services. If legal or any other expert assistance is required, the services of a competent person should be sought. FROM A DECLARATION OF PARTICIPANTS JOINTLY ADOPTED BY A COMMITTEE OF THE AMERICAN BAR ASSOCIATION AND A COMMITTEE OF PUBLISHERS.

LIBRARY OF CONGRESS CATALOGING-IN-PUBLICATION DATA

Available upon Request

ISBN: 978-1-61728-280-5

Published by Nova Science Publishers, Inc. ✦ New York

CONTENTS

Preface		vii
Chapter 1	Introduction	1
Chapter 2	Cases Study	3
Chapter 3	Method	19
Chapter 4	Results and Discussion	35
Chapter 5	Conclusion	47
References		49
Index		53

PREFACE

ZnO nanostructure is a material that is central for many nanotechnology applications, such as chemical and biological sensors. A systematic molecular dynamics study for the behavior of water droplet and electrolyte solutions interacting with ZnO were done. The contact angle of a water droplet on ZnO polar slabs and nanorods/-tubes array changes significantly as a function of the ZnO-water interaction energy and nanostructure geometry. The water contact angle served as a criterion to tune the intermolecular interactions. To recover a hydrophilic surface, voltage range from 1 to 20 volt were applied along the z-axis of the system to simulate the electrowetting behavior case. ZnO nanotube was used to study the permeation of water for equilibrium and applied voltage cases, illustrating the influence of the surface topography and the intermolecular parameters and surface charges on permeation kinetics. We also studied the ionic currents through ZnO nanotubes simulating the case as field effect transistor (FET) of NaCl, KCl, $CaCl_2$, and $MgCl_2$ electrolyte solution for different concentrations (0.1, 0.5, 1.0, 5.0, and 10.0M) and calculate the solution conductance of these ions by applying a voltage difference (1-20V) along the z-axis of the system. We achieved by these molecular dynamics simulations the characteristic behaviors of ZnO nanostructure-electrolyte solution interactions and how it is suitable to use as ion selective sensor for intracellular microenvironment.

Chapter 1

INTRODUCTION

The role of intracellular ions in signaling concentrations and pathways have followed technological breakthroughs in biochemistry, genetics, and developmental biology. Yet the fundamental mechanisms by which the total intracellular ionic state influences the cell response to its chemical and mechanical environment are still undetermined. This deficiency is due in part to limitations of analytical methods for assessing intracellular ion flux in living cells. Many measurements methods rely on the addition of chemical reagents such as ion binding dyes that may alter cell metabolism and provide only limited spatial resolution. Other deficiencies in sensing technologies can be attributed to the difference in size between the molecular processes taking place and the sensor. Recent advances in nanofabrication allow for the creation of new type of sensors. These probes introduce a new methodology controlled insertion into the interior environment of living cells. ZnO nanostructures have recently attracted considerable attention for the detection of chemical ions and biological molecules [1-7]. Among a variety of nano-sensor systems, the ZnO nanostructure electrochemical probe is one that offers high sensitivity and real-time detection. In this work we will try to simulate a probe from a single ZnO nano-rod/-tube or array which capable of electrochemically characterizing the simulated cell interior solutions. The use of a single ZnO nanotube probe will allow for sensing of concentrations and fluxes of electrochemically active ions. A detailed theoretical study of these interactions will be performed using molecular dynamics simulations showing spontaneous and continuous filling of a

ZnO nanotube with ordered chains of water molecules. We will present our results as case study for ZnO hexagonal polar slabs interactions of wetting and electrowetting, wetting and electrowetting of ZnO nanorods/-tubes array, and transport properties of water (filling) and ionic current of (Cl^-, K^+, Na^+, Mg^{2+}, and Ca^{2+}) ions confined in nanoscale one-dimensional channels multi-wall ZnO nanotube which have great interests for physics, biology and material science applications.

Chapter 2

CASES STUDY

The interface of a solid in contact with liquid is of fundamental importance for a wide variety of applications, particularly biosensors. When hydrophobic solids are immersed in liquid, this process is called wetting. The controls of the wetting properties of the sensing surface are important for enhancing the sensitivity and reducing the liquid consumption. One method for modifying the hydrophobicity of a surface is through the application of an external electric field. Such a phenomenon is referred to as electrowetting, where the surface experiences a change in wettability upon the application of an electric field, which initiates a change in the contact angle between liquid-solid systems. There are many factors which affect the performance of an electrowettable surface, such as surface energies and tensions of the materials. The control of the contact angle has been used to drive liquid droplets in micro- or nanosize channels on biochemical and environmental sensing devices. In such applications, a faster contact angle transition rate is desired for prompt control of liquid movement. Limited work has been done to understand how different surface structures and morphologies affect the surface wettability of ZnO nanostructures.

2.1. ZnO Hexagonal Polar Surfaces Slab-Water Interaction (Wetting and Electrowetting)

When a liquid is in contact with an inert solid phase, the liquid "wets" the surface. Liquid molecules at the solid-liquid interface are now in a different environment than the ones that are either in the bulk or at the exposed surface. Those molecules feel two kinds of forces: cohesive forces acting between like molecules and adhesive forces acting between different molecules. The balance between cohesive and adhesive forces determines the wetting properties of the surface. When the cohesive forces of water can be counterbalanced by the adhesive forces of the substrate, a liquid droplet tends to spreads over the surface. When the cohesive forces of water are stronger than the adhesive forces of the substrate, the droplet tries to avoid the surface, keeping its spherical shape and reducing the surface tension.

For a perfect homogeneous, horizontal, isotropic solid surface the contact angle of a liquid droplet (sufficiently large so that the effects of the three-phase contact line can be neglected) is obtained by the Young's equation by balancing the horizontal component of the forces acting on the three-phase (solid-liquid-vapor) contact line [8]:

$$\gamma_{sv} = \gamma_{sl} + \gamma_{lv} \cos\theta_Y \qquad (1)$$

$$\cos\theta_Y = \frac{\gamma_{sv} - \gamma_{sl}}{\gamma_{lv}} \qquad (2)$$

where θ_Y is Young contact angle and γ_{ij} is the surface tension or surface free energy, respectively. The indices stand for the different interfaces: solid-vapor (*sv*), solid-liquid (*sl*), and liquid-vapor (*lv*). In the present context, when characterizing the interactions of solid surfaces with water, we refer to surfaces with $\theta_Y > 90°$ as hydrophobic and those with $\theta_Y < 90°$ as hydrophilic.

Young's equation is strictly valid only for macroscopic droplets at mechanical equilibrium with any adsorbed film of thickness l (see figure 1) on a molecularly smooth horizontal solid surface.

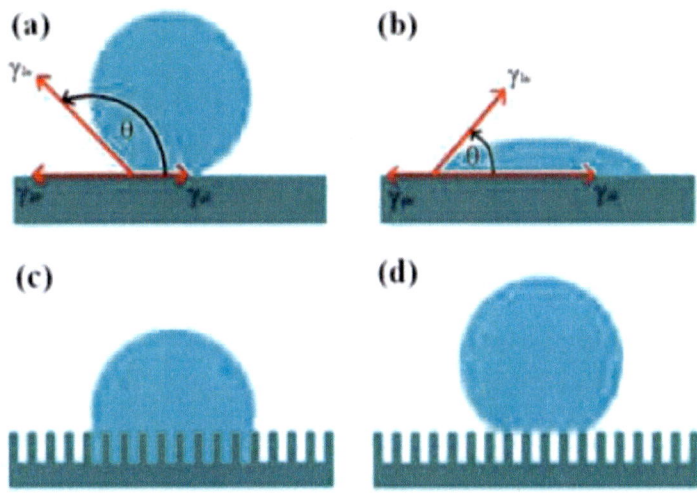

Figure 1. A small droplet in equilibrium over a horizontal surface with different degree of wetting: (a) the wetting is low and the contact angle is large, (b) the wetting is high and the contact angle is small. Droplets wetting rough surfaces: (c) Wenzel state with enhanced solid-liquid interfacial area, (d) the Cassie-Baxter state with entrapped air underneath the droplet (non-wetting condition).

For microscopic droplets the contact angle will be influenced by surface interactions and the nature of the three-phase solid-liquid-vapor contact line of length L_{slv}, which will contribute an additional free energy per unit length or line tension τ to the excess free energy of the droplet. This effect adds an excess energy term $\tau \, dL_{slv}$ to Eq. 1. Hence for a spherical droplet of contact angle θ and lateral radius r_B, we obtain the modified Young's equation [9]:

$$\gamma_{sv} = \gamma_{sl} + \gamma_{lv} \cos\theta_Y + \frac{\tau}{r_B} \qquad (3)$$

which can alternatively be written as

$$\cos\theta_w = \cos\theta_Y - \frac{\tau}{r_B \, \gamma_{lv}} \qquad (4)$$

where θ_Y is the contact angle in the limit of very large droplets $r_B \to \infty$. Hence from Eq. 4 the quantities τ and $\cos\theta_Y$ can be determined from a study of $\cos\theta_w$ as a function of $1/r_B$.

The contact angle θ_Y provides important information about the wettability of a surface. A liquid partially wets a surface if a large droplet possesses a finite contact angle, $\theta_Y > 0$. If, however, this liquid has a contact angle $\theta_Y = 0$ the liquid completely wets the surface and the surface is covered by a thick liquid film.

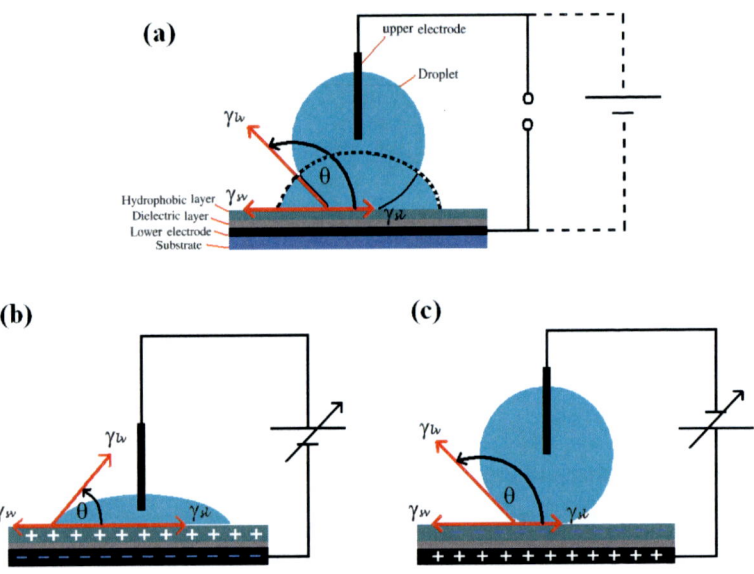

Figure 2. Electrowetting set-up; (a) partially wetting liquid droplet at zero voltage (solid line) and at high voltage (dash line). Operation of voltage between the droplet and the electrode (b) forward biasing voltage which changes the distribution of charges due to dielectric insulator and decreases the contact angle , (c) backward biasing voltage which reverse the surface polarity (hydrophobic surface) and increases the contact angle.

In electrowetting, generally we are dealing with droplets of partially wetting liquids on planar solid surface substrates where in most interested applications, the droplets are aqueous salt solutions with a typical size of the order of 1mm or less. In case of applying external electric field the electrowetting for homogeneous planar solid substrates can be described by Lippmann's equation of electrowetting which is based on general Gibbsian interfacial thermodynamics (see figure 2) [10].

Surface wettability can be enhanced by application of electric voltage driving dipolar water molecules to the field-exposed region. Upon applying a voltage dV, an electric double layer builds up spontaneously at the solid-liquid interface consisting of charges on the metal surface on the one hand and a cloud of oppositely charged counter-ions on the liquid side of the interface. Since the accumulation is a spontaneous process, the effective interfacial tension of solid-liquid can be defined as:

$$d\gamma_{sl}^{eff} = -\sigma_{sl} dV \qquad (5)$$

where $\sigma_{sl} = \sigma_{sl}(V)$ is the surface charge density of the counter ions. The voltage dependence of γ_{sl}^{eff} is calculated by integrating the above equation. In general, this integral requires additional knowledge about the voltage dependent distribution of the counter ions near the interface which is calculated on the basis of the Poisson-Boltzmann distribution. For the simplicity we assume that the counter-ions are all located at a fixed distance (in order of a few nm) d_H from the surface (Helmholtz model). In this case, the double layer has a fixed capacitance per unit area,

$$C_H = \frac{\varepsilon_o \varepsilon_l}{d_H} \qquad (6)$$

where ε_l is the dielectric constant of the liquid.

We obtain [11]:

$$\gamma_{sl}^{eff}(V) = \gamma_{sl} - \frac{\varepsilon_o \varepsilon_l}{2d_H}(V - V_{pzc}) \tag{7}$$

where V_{pzc} is the potential difference of zero charge (metal surfaces acquire a spontaneous charge when immersed into electrolyte solution at zero voltage and V_{pzc} is the voltage required to compensate for this spontaneous charging), and γ_{sl} the chemical contribution to the interfacial energy is assumed to be independent of the applied voltage.

To obtain the response of the contact angle for an electrolyte droplet placed directly on an electrode surface we used this equation [12]:

$$\cos\theta_w = \cos\theta_Y + \frac{\varepsilon_o \varepsilon_l}{2 d_H \gamma_{lv}}(V - V_{pzc}) \tag{8}$$

For typical values of $d_H = 2\,nm$, $\varepsilon_l = 81$, and $\gamma_{lv} = 72\,dyne/cm$ we find that the ration on the term is on the order of $1V^{-2}$. The contact angle thus decreases rapidly upon the application of a voltage.

To include the effect of the nanometer size of the water droplet on the WCA the Eq. 8 will be as follow:

$$\cos\theta_w = \cos\theta_Y - \frac{\tau}{r_B \gamma_{lv}} + \frac{\varepsilon_o \varepsilon_l}{2 d_H \gamma_{lv}}(V - V_{pzc}) \tag{9}$$

When semiconductor material get in contact with liquid depletion will create which insulates the droplet from the electrode and the electric double layer builds up at the depletion-droplet interface. Since the depletion layer thickness d is usually much larger than d_H, the total capacitance of the system is reduced tremendously. The system can be described as two capacitors in series, namely the Helmholtz capacitor C_H and the semiconductor layer (depletion layer) C_d defined as:

$$C_d = \frac{\varepsilon_o \varepsilon_d}{d} \tag{10}$$

Since $C_d \ll C_H$, the total capacitance per unit area will be $C \approx C_d$. With this approximation, we neglect the finite penetration of the electric field into the liquid. As a result, we find that the voltage drop occurs within the depletion layer and the Eq. 7 is replaced by [11,12]:

$$\gamma_{sl}^{eff}(V) = \gamma_{sl} - \frac{\varepsilon_o \varepsilon_l}{2d} V^2 \qquad (11)$$

Here we assume that the surface of the depletion layer does not give rise to spontaneous adsorption of charge in the absence of an applied voltage, i.e., we set $V_{pzc} = 0$. We can rewrite the above equation in term of contact angle as:

$$\cos\theta_w = \cos\theta_Y + \frac{\varepsilon_o \varepsilon_l}{2d\gamma_{lv}} V^2 \qquad (12)$$

Then we add the nano-droplet modified term the above equation will be:

$$\cos\theta_w = \cos\theta_Y - \frac{\tau}{r_B \gamma_{lv}} + \frac{\varepsilon_o \varepsilon_l}{2d\gamma_{lv}} V^2 \qquad (13)$$

2.2. ZnO Nanorods or Tubes Array-Water Interaction (Wetting and Electrowetting)

There have been a lot of interests in studying the wetting behaviors of ZnO nano-rods/-tubes array, which is a technologically important for many promising applications in the fields of chemical and biological sensing where the surface wettability plays a very important role. These nanostructures were generally used to enhance the surface wettability of ZnO films and, in some cases, to obtain superhydrophobicity surfaces.

The wetting properties of the nanostructures are different from the smooth surfaces. The principles of wetting rough surfaces have been investigated by Wenzel [13] with the assumption that the entire surface under the drop is covered with liquid (see figure 1c). The Young's equation is modified as:

$$\cos\theta_w = \alpha \frac{\gamma_{sv} - \gamma_{sl}}{\gamma_{lv}} = \alpha \cos\theta_Y \qquad (14)$$

where θ_w is the contact angle predicted by Wenzel's model, θ_Y is the Young contact angle and α is the roughness factor, defined as the ratio of the actual area of a rough surface to the geometric projected areas.

For a rough hydrophobic surface however liquid may not completely penetrate into surface hollows and trapped air inside, forming a composite solid-air-liquid interface. Cassie and Baxter extended Wenzel's equation to describe the contact angle on such a surface (see figure 1d) [14]:

$$\cos\theta_w = f_1 \cos\theta_1 + f_2 \cos\theta_2 \qquad (15)$$

Where f_1 and f_2 are the area fractions of liquid-solid interface and liquid-air interface, respectively. As $f_1 + f_2 = 1$ and $\theta_2 = 180^o$ due to the non-wetting condition with air, the equation above can be reduced to:

$$\cos\theta_w = f_1 \cos\theta_1 - f_2 \qquad (16)$$

$$\cos\theta_w = f_1(\cos\theta_y + 1) - 1. \qquad (17)$$

This is in agreement with the fact that surface hydrophobicity improves when there is more air trapped between the liquid and solid surfaces.

We can idealize the ZnO nanorods as flat-top, hexagonal sectioned rods with a height of h and a top radius of the rod of R, as illustrated in figure 3. When a water droplet contacts the surface, the area fraction of liquid-solid interface f_1 can be determined by [15]:

$$f_1 = \frac{\left(\dfrac{3\sqrt{3}}{2}R^2\right) + 6Rh'}{\dfrac{A}{n} + 6Rh} \tag{18}$$

where A is the projected area of the water droplet on the ZnO nanorod film surface, n is the number of nanorods in the area A ($A/n \geq (3\sqrt{3}/2)R^2$), and h' ($h' \leq h$) is the depth that water intrudes into the gaps between adjacent rods. When adjacent nanorods are closer (n increases) in a certain area A, the depth of intruding water h' will decrease. The decrease of h' will diminish the increase of f_1 caused by the increase of n and may even lead to a net decrease of f_1.

For a rough hydrophobic surface however we can use the Cassie and Baxter extended equation to describe the contact angle on such a surface in case of electrowetting and to include the effect of the nanometer size of the water droplet on the WCA, we modify Eq. 9 to become:

$$\cos\theta_w = [f_1(\cos\theta_Y + 1) - 1] - \frac{\tau}{r_B \gamma_{lv}} + \frac{\varepsilon_o \varepsilon_l}{2d_H \gamma_{lv}}(V - V_{pzc})^2 \tag{19}$$

We can call the extended equation as Lippmann-Cassie and Baxter electrowetting equation for nano-structure surfaces.

For the case where $V_{pzc} = 0$, the final equation will be:

$$\cos\theta_w = [f_1(\cos\theta_Y + 1) - 1] - \frac{\tau}{r_B \gamma_{lv}} + \frac{\varepsilon_o \varepsilon_l}{2d_H \gamma_{lv}}V^2 \tag{20}$$

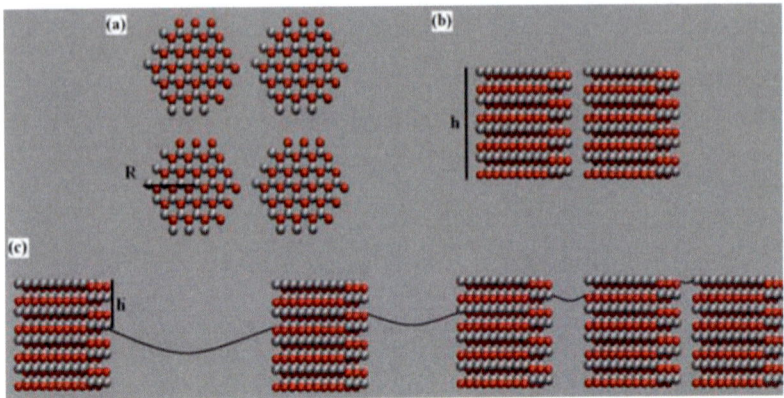

Figure 3. Schematic diagram of roughness factor calculation; (a) and (b) ZnO nanorods array representation which composite of a certain number of ZnO rods n of radius R and length h in area A. (c) Depth of the intruding water h' will decrease with increasing n (adjacent nanorods are closer).

2.3. WATER PERMEATION THROUGH ZNO NANOTUBE

Transport of water into and out of nanotube will be studied in this case. The geometry of a tube (its inner radius, length and shape) and the chemical character of the tube inner wall have a great influence on the permeation of water (see figure 4). We have adopted the Oliver et al. [16] model for permeation of water and ions through nano-pores. This approach model is used to label the two phase states that described the water permeation exhibits in the simulation. We either find liquid-filled tube or vapor-filled one. Due to the small tube inner volumes "vapor" typically refers to zero or one to water molecules in the cavity. Where the water density in the tube is used as an indicator of the phase state, when the water density $b(t)$ rises above 0.65 of the density of bulk water, b_o ($b_o = 1.0\ g\ cm^{-3}$ at $T = 300\ K$ and $P = 1\ bar$) the liquid state is assigned to the phase state at time t, or when $b(t)$ drops below $0.25\ b_o$ the vapor state is assigned.

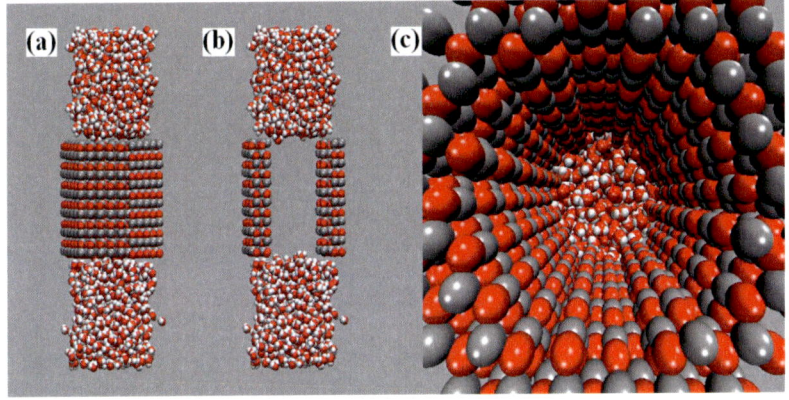

Figure 4. Permeation of water through a ZnO nanotube initial MD set-up; different representations of the system (a) side view, (b) y-axis cross section of the system, (c) inner view through the ZnO tube.

The model assumes the inner tube volume is $\pi r^2 L$, and this volume can exchange water molecules with the bulk water outside the tube, which acts as a particle reservoir at average chemical potential μ. The free energy that describe the tube in the closed and open state used as $\nabla U(T,V,\mu) = -PV$. Then free energy difference between the vapor and the liquid state with the corresponding surface contributions will define as:

$$\Delta U(r) = 2\left[\gamma_{lv} + \frac{1}{2}\Delta\mu\,\Delta b_{lv}\,L\right]\pi r^2 + 2\pi L \Delta\gamma_s\,r \qquad (21)$$

where $\Delta\mu = \mu - \mu_{sat}$ is the distance of the state from saturation, $\Delta b_{vl} = b_l - b_v$ the difference in densities at saturation, and $\Delta\gamma_s = \gamma_{vs} - \gamma_{ls}$ the difference in surface free energies of the two phases with the wall. The term $\Delta\mu\,\Delta b_{vl}\,L$ is small for $L < 10\,nm$ as the system is close to phase coexistence (for $L \approx 1\,nm$ it is about 10^{-3} times smaller than $\gamma_{lv} = 17\,k_B\,T\,nm^{-2}$ when estimated from

$$\Delta b_{vl}\,\Delta\mu\,L \approx \Delta P\,L \approx 1\,bar \times L = 2.4\times 10^{-3}\,k_B\,T\,nm^{-2} \times L,\ at\ T = 300K$$

and will be neglected. Only the difference between the surface free energies enters the model so we express it as the contact angle θ_Y, using the macroscopic definition from Young's equation [17]. Then Eq. 21 becomes:

$$\Delta U(r, L, \theta_Y) = 2\pi r \gamma_{lv}(r + L\cos\theta_Y) \qquad (22)$$

For fixed tube length L and a given tube material, characterized by θ_Y, the graph of $U(r)$ over the tube inner radius r describes a parabola containing the origin [18].

The connection between the model free energy $U(r)$ and the behavior of the system as observed in MD simulations, it is assumed that g_i as numbers of equilibrium discrete states (MD equilibrium trajectory), i.e. the openness $\langle g(r) \rangle$, is defined as

$$\langle g(r) \rangle = \frac{1}{1 + \exp\left(-\frac{1}{k_B T}\Delta U(r)\right)} \qquad (23)$$

The macroscopic relation due to Young and Lippmann can describe electrocapillarity in a planar confinement as (see figure 5) [19]:

$$\cos\theta_w = \cos\theta_Y - \frac{W_{el}(V)}{2\gamma_{lv}} = \cos\theta_Y + \frac{CV^2}{2\gamma_{lv}} \qquad (24)$$

Here $W_{el}(V)$ is the change in electrostatic free energy per unit area, associated with surface spreading of the liquid, wetting both walls (hence the factor 1/2), V is the voltage across the interface, and θ_Y the contact angle in the absence of electric field. The form of W_{el} depends on system geometry and material properties but is generally presumed to be proportional to the areal electric capacitance of the interface, C, and the potential drop across the interface squared.

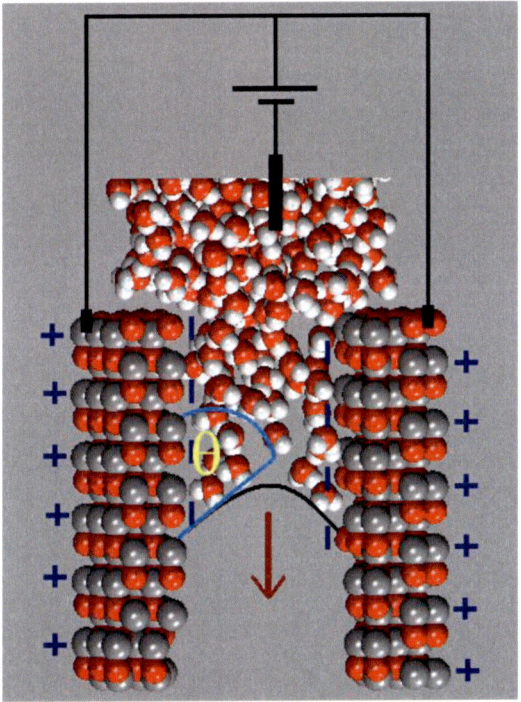

Figure 5. Water Electro-permeation through ZnO nanotube. In this case the ZnO tube walls act as a dielectric insulator.

2.4. IONIC CURRENTS OF MG^{2+}, CA^{2+}, K^+, AND NA^+ IONS THROUGH ZNO NANOTUBE

To calculate the current-voltage characteristics of the electrolyte we will follow the mobility model based on molecular dynamics would generalize this approximation. The potential variation through the nanotube with the following analytical expression, which has been shown to be valid from molecular dynamics in nanotubes [20]:

$$V(z) = \frac{V_0}{\pi} \tan^{-1}\left(\frac{z}{L_{eff}}\right) \qquad (25)$$

where V_o is the external voltage across the device driving the ions through the nanotube and L_{eff} is a characteristic length (not the channel length) so that the potential achieves its electrode values at $z = [-L_z/2, L_z/2]$ where L_z is the channel length, with non-zero electric field at the electrodes. We assume the following approximation that reservoir resistance is negligible in comparison with nanotube resistance and the potential in the reservoir is constant.

Considering both anion density current J_a and cation density current J_c and neglecting the diffusion current, the current density for each ion type is given by [20,21]:

$$J_a = q\mu_a c_a \nabla \psi \qquad (26)$$

$$J_c = q\mu_c c_c \nabla \psi \qquad (27)$$

And the total current density is:

$$J = J_a + J_c \qquad (28)$$

where ψ is the electrostatic potential, c_a and c_c are the anion and cation concentration, μ_a and μ_c are the anion and cation mobility, respectively. Assuming no recombination inside the tube, we have [20]:

$$\nabla J_{a,c} = 0 \qquad (29)$$

which by using the divergence theorem, implies that the current is constant through the nanotube. Due to the one dimensional nature of the external potential (Eq. 25), this condition (Eq. 29) is not fully satisfied. Therefore, we spatially average the current through the nanotube to eliminate the slight J-variations due to the slanted geometry of the nanotube [20]:

$$\langle I_{a,c} \rangle = \frac{1}{L} \int_{L/2}^{-L/2} dz \int \int_{S(z)} J_{a,c}(r) dS \qquad (30)$$

Here L is the length of the tube and $S(z)$ is the nanotube cross section at ordinate z. In this context, we define each ion conductance in the nanotube as [20]:

$$G_{a,c} = \frac{\langle I_{a,c} \rangle}{V_o} \qquad (31)$$

Chapter 3

METHOD

To study the sensitivity of the contact angle to the ZnO-water interaction potential and the droplet size and shape, a series of MD simulations of water droplets on ZnO nanostructures are performed. The MD simulation techniques are described along with the details on how the contact angles, density profiles, and line tension are extracted from the simulations. In these cases, we will describe how to build atomic-scale models of ZnO hexagonal polar surface slabs and ZnO nano-rod/-tube arrays using computational procedures that mimic the surface wetting experiment. The shape and dimensions including diameter size for the water droplet were varied. The surface morphology and dimensions including diameter size, rod length, and density were varied. The surface wettability of ZnO films were examined by water contact angle measurements. Switchable wettability was also investigated on the both types of films (rods and tubes) by applying voltage difference across the system. Water permeation through ZnO nanotube was studied by varying the inner radius of the nanotube for different applied voltages. Ionic current through ZnO nanotube were studied by applying voltage along the z-axis of the system with varying the salt concentration of the electrolytes for NaCl, KCl, $CaCl_2$, and $MgCl_2$ solutions.

3.1. MOLECULAR DYNAMICS

Molecular dynamics (MD) simulations can be used to complement the experimental data on the ZnO–water interaction. Since interactions with biomolecules in all biotechnical applications occur in an aqueous environment, an empirical force field for ZnO should be able to reproduce its surface wetting properties. For the ZnO-water dynamics we used the molecular dynamics program NAMD 2.8 [22] and the simulation outcomes were analyzed through our own routines programmed in Matlab 7.5 [23] and VMD 1.8.7 [24] programs. The universal force field (UFF) provided by Rappé et al. [25] were employed to produce hexagonal polar slab, nanorod and nanotube from ZnO. The UFF is a harmonic force field, in which the total potential energy expression is expressed as:

$$E = E_R + E_\theta + E_\phi + E_w + E_{vdW} + E_{el} \tag{32}$$

where E_R is the bond stretching energy, E_θ is the bond bending energy, E_ϕ is the torsions energy, E_w is the inversions energy, E_{vdW} is the ven der Waals interaction energy and E_{el} is the electrostatic interaction energy. The functional form of the above energy terms are given as follows:

$$E_R = k_1(r - r_o)^2 \tag{33a}$$

$$E_\theta = k_2(A_o + A_1 \cos\theta + A_2 \cos 2\theta), \text{ where} \tag{33b}$$

$$A_2 = \frac{1}{(2\sin^2\theta_o)},$$

$$A_1 = -4A_2 \cos\theta_o,$$

$$A_o = A_2(2\cos^2\theta_o + 1),$$

$$E_\phi = k_3(1 \pm \cos n\phi) \quad (33c)$$

$$E_\omega = k_4[1 + \cos(n\chi - \chi_o)] \quad (33d)$$

$$E_{vdW} = D\left[\left(\frac{r^*}{r}\right)^{12} - 2\left(\frac{r^*}{r}\right)^6\right] \quad (33e)$$

$$E_{el} = \frac{q_i q_j}{\varepsilon r_{ij}} \quad (33f)$$

where k_1, k_2, k_3, and k_4 are force constants, θ_o is the natural bond angle, D is the van der Waals well depth, r^* is the van der Waals length, q_i is the net charge of an atom, ε is the dielectric constant, r_{ij} is the distance between two atoms. In the calculation, the difference of the non-polar and polar semiconductors is dependent of the net charge q_i being set to zero or not.

To parameterize a force field that can account for ZnO-water interactions we employed a potential energy function compatible with the CHARMM force field [26]. We began with previously existing UFF parameters for ZnO and refined them to reproduce the interactions of ZnO with water. The functional form used is:

$$E_{Total} = E_{Bond} + E_{Angle} + E_{vdW} + E_{Coulomb} \quad (34)$$

The first two terms on the right-hand side of Eq. 34 are harmonic potentials used to describe bond stretching and bending:

$$E_{bond} = \sum_{bonds\,i} k_i^{bond}(r_i - r_{0i})^2 \quad (35)$$

$$E_{angle} = \sum_{angles\,i} k_i^{angle}(\theta_i - \theta_{oi})^2 \quad (36)$$

Here the sums run over all bonds and bond angles; the parameters k_i^{Bond}, k_i^{Angle}, r_{oi}, and θ_{oi} describe the equilibrium values of the degrees of freedom. The bonded interactions of ZnO were taken from UFF but adjusted to fit Eqs. 35 and 36.

The last two terms in Eq. 34 describe the vdW and electrostatic nonbonded interactions that are the main focus of this work:

$$E_{vdW} = \sum_i \sum_{j>i} D_{ij} \left[\left(\frac{r_{ij}^{min}}{r_{ij}} \right)^{12} - 2 \left(\frac{r_{ij}^{min}}{r_{ij}} \right)^{6} \right] \quad (37)$$

$$E_{Coulomb} = \sum_i \sum_{j>i} \frac{q_i q_j}{4\pi\varepsilon_o r_{ij}} \quad (38)$$

The vdW interactions are represented through a Lennard-Jones 6-12 potential, with two adjustable parameters: r_{ij}^{min}, the distance at which the energy between atoms i and j is minimal, and D_{ij}, the well depth. In the case of the oxygen atom, vdW parameters were taken from the CHARMM force field, assigning 3.5 Å to r_O^{min} and 0.15 kcal/mol to D_O. For zinc, r_{Zn}^{min} was set to 2.763 Å, and 0.25 kcal/mol to D_{Zn} corresponding to the zinc atomic radius [25]. The electrostatic interactions were calculated using Eq. 38, q_i and q_j being the partial atomic charges of atoms i and j, and ε_o being the vacuum permittivity. The surface charges corresponding to zinc q_{Zn} and oxygen q_O were tuned to reproduce the ZnO wettability. We employed the TIP3P model of water, since the CHARMM force field works best with this choice of model and its functional form conforms to Eq. 34.

3.2. BUILDING ZnO STRUCTURES

3.2.1. ZnO Hexagonal Polar Slab

ZnO is polar semiconductor; his polarity came from bulk characteristic of hexagonal wurtzite materials that stems from the lack of a symmetry center and results in two inequivalent c-directions. The noncentrosymmetric structure and the partial ionicity of the bonds, which results from the differing Pauling electronegativity of the atoms, yield a net spontaneous polarization field along the c-axis. Therefore, the Zn-polar ZnO (0001) and O-polar ZnO ($000\bar{1}$) are characterized by different properties, such as morphology, dipole moment, surface charge, band bending, and chemical stability. Most significantly, a strong polarization-induced surface charge of opposing sign characterizes the (0001) and the ($000\bar{1}$) polar surfaces. To generate the ZnO hexagonal polar slab structures we used the plugin of VMD program. The schematic procedure for building ZnO nanostructure is shown in figure 6. To create hexagonal polar slab of ZnO, we replicated the unit cell with lattice parameters a=3.249 Å and c=5.207 Å [27] for 26-26-2 times and yielding a system of 5408 atoms. This step was followed by the cut the slab to hexagonal shape having 4056 atoms with radius 42.237 Å and slab thickness 10.414 Å (see figure 6d).

3.2.2. ZnO Nanorods

To generate the ZnO nanorods and nanotubes array structures we used the VMD program. The schematic procedure for building ZnO nanostructure is shown in figure 6b. To build hexagonal rod of ZnO, we replicating the ZnO unit cell for 6-6-4 times and yielding a system of 576 atoms. Subsequently, the bulk system was cut in hexagonal shape to reproduce ZnO hexagonal rod crystal having 432 atoms with dimensions (r=9.747 Å and l_z=20.828 Å).

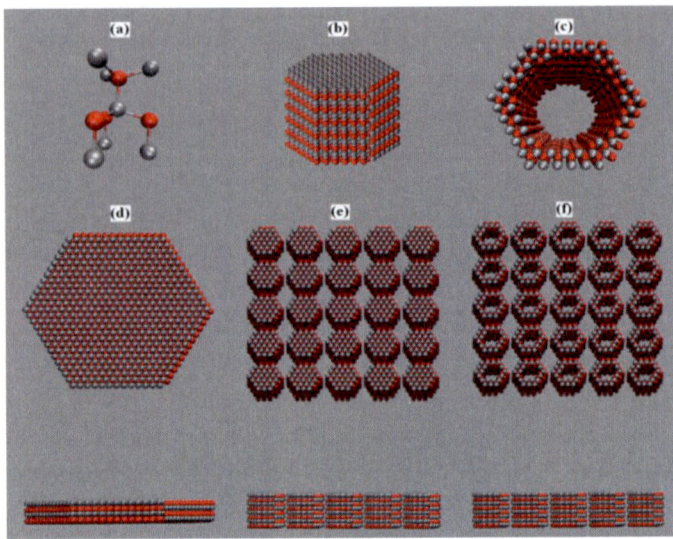

Figure 6. Building steps of ZnO nano-structures. (a) Unit cell of ZnO, which is replicated to generate the hexagonal shape of ZnO nanorod (b), the cylinder inner shape drilled to make the ZnO nanotube (c), a thin hexagonal slab of the ZnO polar surfaces were created by replicating the unit cell 26-26-2 times (d). ZnO nanorods/tubes array were created by replicating the ZnO rod/tube at equal distances in x- and y-axes, (e) for rods array and (f) for tubes array.

3.2.3. ZnO Nanotube

To model the ZnO nanotube, a hexagonal prism tube of ZnO, with length of 2.0828 nm, was constructed by removing atoms from the hexagonal ZnO nanorod structure; we produce a cylindrical tube of symmetric shape with inner radius 0.81225 nm. The length of the cylinder tubes equals the hexagonal ZnO nanotube length (see figure 6c). However, the detailed shape of the cylinder tube surface is not fully regular and the presence of irregularities and deviations from a symmetrical cylinder structure are considered. These irregularities can affect the motion of water molecules or ions through the ZnO nanotube. The total charge of a nanotube was found to approximately be zero.

To make ZnO nanorods/tubes array we replicate the nanorod/tube structures at equal distance (in x and y directions) from each other's and we have varied these distance to change the density of nanorods/tubes in the array (in our case 2.2 nm center-center distance with rods or tubes density per area equal to 0.65 nm^{-2}) (see figure 6e and 6f).

3.3. ZNO-WATER SYSTEMS

3.3.1. Wetting

To simulate the water contact angle (WCA) of ZnO [28,29]; rectangular and spherical volumes of water were generated from a pre-equilibrated water box using the solvate plugin of VMD. A water box with side lengths 30-30-30 $Å^3$, and 40-38-30 $Å^3$ composed of 2478, and 4236 atoms and a water sphere with radius of 16 Å composed of 1428 atoms were placed on slab top of the ZnO hexagonal polar surfaces (see figure 7a,b). To perform MD simulations with full electrostatics under periodic boundary conditions, the total charge of the system has to be zero. To maintain electroneutrality of the systems for ZnO-water wetting interaction, the surface atoms were assumed as follows: two atoms were considered covalently bonded if they had a separation distance of 2 Å or less; oxygens with four bonds and zinc with four bonds were classified as a non-dangling type; oxygens with less than four bonds and zinc with less than four bonds were classified as a dangling oxygen type and dangling zinc type, respectively, or just as a dangling atom type.

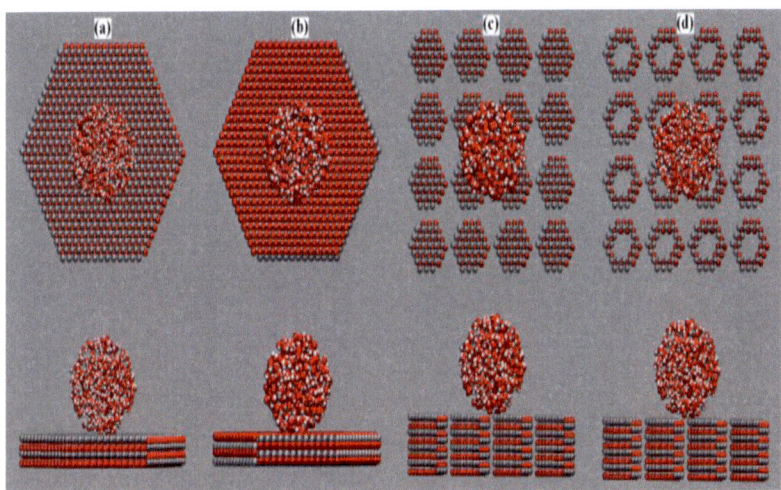

Figure 7. Top view and side view of the initial (t=0) water droplet (sphere) on Zn- polar slab (a), O- polar slab (b), ZnO nanorods array (c), and ZnO nanotubes array (d).

All simulations involving ZnO-water systems were carried out using a CHARMM compatible form of the potential energy function (Eq. 34). Simulations were performed with an integration time step of 1 fs for 1ps using the conjugate gradient method and then equilibrated for 2 ns in the *NVT* ensemble for periodic boundary conditions at 300 K, vdW interactions were calculated with a cutoff of 12 Å (switching function starting at 10 Å), and the long-range electrostatic forces were calculated using PME summation method. A Langevin thermostat was used to maintain a constant temperature in the NVT ensemble simulations. Simulations in the NpT ensemble were performed using a hybrid Nose-Hoover Langevin piston. To keep the ZnO structure rigid, the Zn and O atoms were either fixed or restrained to their original position by applying a harmonic force with a force constant of 10 kcal/mol/Å². The water temperature remains stable during the MD runs; it has an average of 298.9 K and a standard deviation of 5.0 K. Figure 8 illustrates the initial and equilibrate configuration. Samples of the trajectory are stored every 0.1 ps.

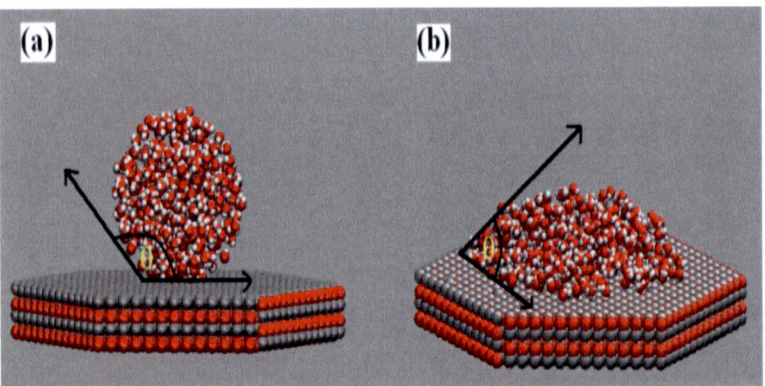

Figure 8. Side view of the initial (t=0) (a) and equilibrated (t=0.4 ns) (b) water droplet (sphere) on (0001)-Zn polar hexagonal slab.

3.3.2. Electrowetting

To simulate the electrowetting of the ZnO nanostructures, we applied a uniform electrical field to all atoms; it induces, at the beginning of the simulation, a rearrangement of the water molecules. The resulting voltage bias V across the simulated system depends on both the magnitude of the applied field E and the dimension L_z of the system in the direction of the field, i.e. $V = E L_z$ [30]. A constant electric field was applied to produce the desired drifting voltage across the system. The bias voltage was applied along (4.2 nm for the case of ZnO slab-water sphere, 4.0 nm for the case of ZnO slab-water box, 5.2 nm for the case of ZnO rod or tube-water sphere, and 5.0 nm for the case of ZnO rod or tube-water box) system length in z-direction (solution - ZnO nano structures) (see figure 9).

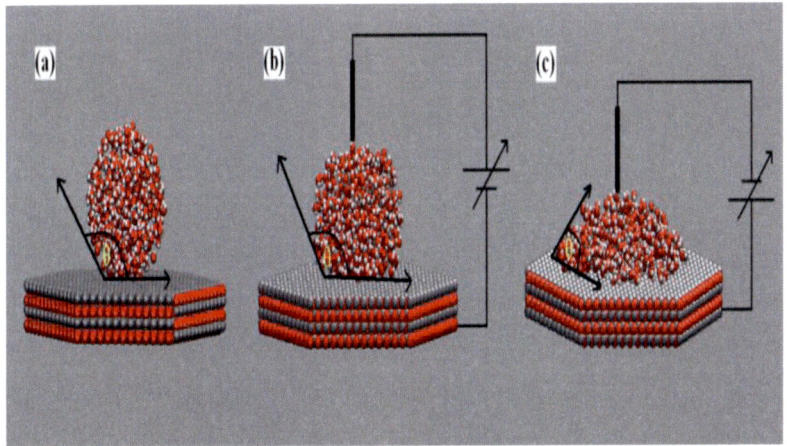

Figure 9. Side view of the initial (t=0) (a) and electro-wetted (t=0.6 ns) water droplet sphere on the (0001)-Zn polar slab at 20 V in the forward biasing (b) and -20 V in the backward biasing (c).

For the case of ZnO nanorods/tubes array we also approximate the problem to four nanorods/tubes system array to simulate the electrowetting to reduce the time of simulation, a water box with 40-38-30 $Å^3$ composed of 1412 water molecules was placed on top of four nanorods or nanotubes array (see figure 10). All systems were minimized for 1ps with time step 1 fs using the conjugate gradient method and then equilibrated for 2 ns in the NVT ensemble. To keep the ZnO nanorod/tube rigid, the Zn and O atoms were either fixed or restrained to their original position by applying a harmonic force with a force constant of 10 kcal/mol/$Å^2$. Figure 11 illustrates the before and after the applied voltage. Samples of the trajectory are stored every 0.1 ps.

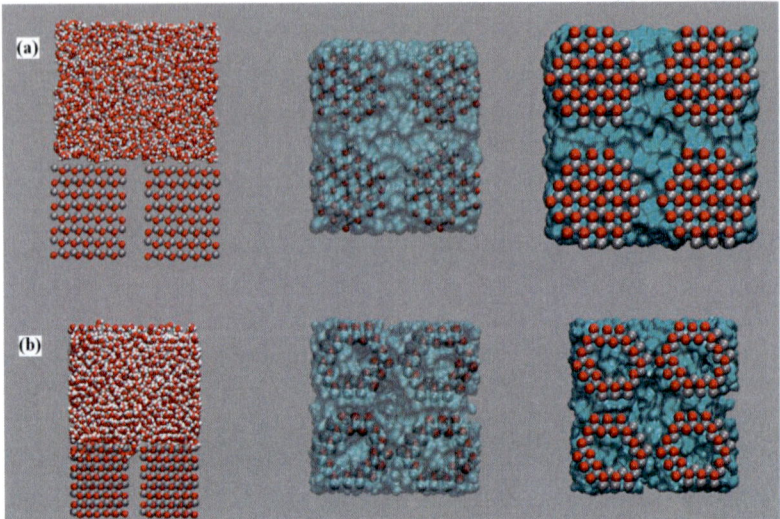

Figure 10. Initial array wetting systems with different viewing sides, consist of four ZnO nanorods array (a) or four ZnO nanotubes array (b).

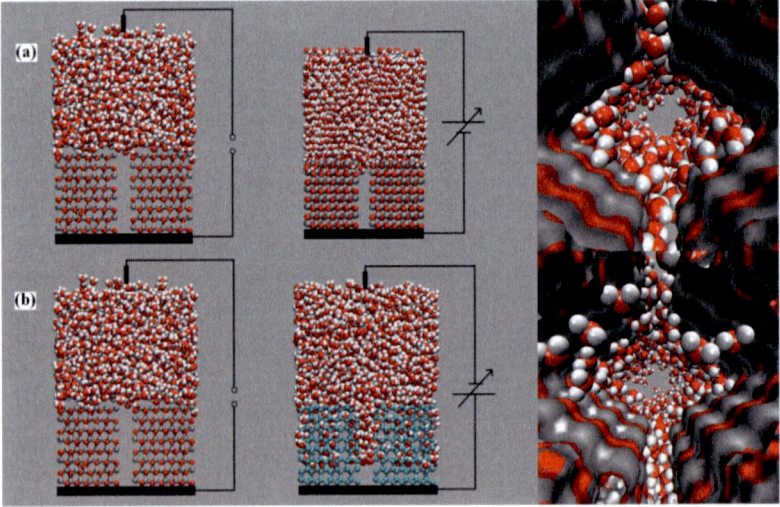

Figure 11. 20 V forward biasing for electrowetting of four ZnO nanorods array with different views (a) and 20 V backward biasing for four ZnO nanotubes array with different views (b), respectively. Nanorods case shows non-wetting behavior where in case of nanotubes it shows wetting behavior.

3.3.3. Water Permeation

We simulated the permeation of water through the ZnO nanotubes using the intermolecular parameters validated by our simulations. The ZnO nanotube was covered with a water box (20-20-30 Å^3) on each side then we cut the two boxes to hexagonal shape to allow us to use hexagonal periodic boundary conditions, and the resulting systems (having l_z=9.1nm) were equilibrated for 1 ns at 300 K under different conditions. The number of water molecules within the nanotube was used as a measure of permeation (see figure 12).

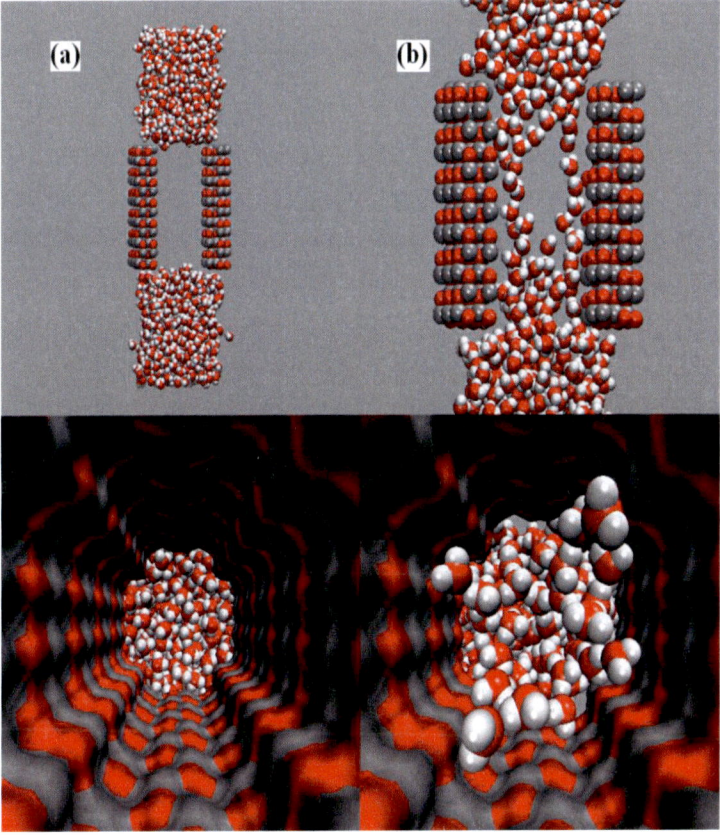

Figure 12. Initial at t=0 (a) and equilibrated at t=0.4 ns (b) water permeation through ZnO nanotube at T=300 K.

A large number of simulations were run for water in model tubes order to investigate the influence of inner tube radius (0.12 nm ≤ R ≤ 1 nm) and inner tube surface character (hydrophobic vs. hydrophilic). To count the number of water molecules in the ZnO nanotube, we considered a cylindrical bin of 32 Å diameters concentric with the tube axis, with a height of 32 Å parallel to the Z-axis.

3.3.4. Ionic Currents

The simulation was performed for ZnO nanotube in presence of water molecules in contact with both top and base sides of the nanotube. The water is extended ~3 nm above and below the ZnO nanotube. Ions including Na^+, K^+, Ca^{2+}, Mg^{2+} and Cl^- were added to obtain a 10.0, 5.0, 1.0, 0.5, 0.1M sodium chloride (NaCl) or potassium chloride (KCl) or calcium chloride ($CaCl_2$) or magnesium chloride ($MgCl_2$) solutions (see figure 13).

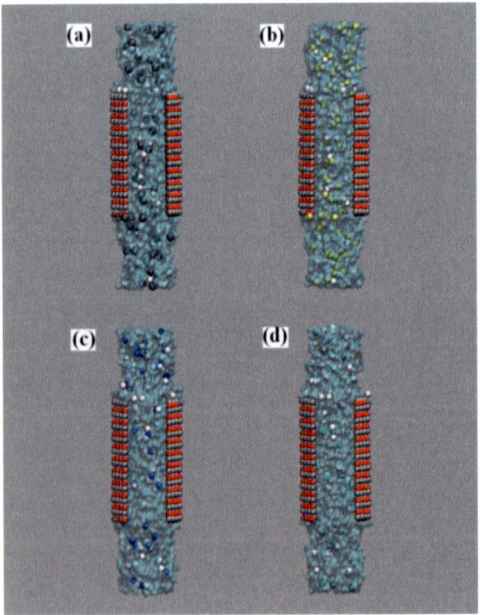

Figure 13. Filled ZnO nanotube with 5M concentration of (a) NaCl, (b) KCl, (c) $CaCl_2$, and (d) $MgCl_2$ electrolyte solutions.

Method

We have used the molecular dynamic simulations to perform ions transition through the ZnO nanotubes. To obtain a relative permittivity of 8.7 [31], harmonic restraints were applied to bulk ZnO atoms and harmonic bonds were applied between neighboring atoms. Similar forces were applied to surface atoms. In all simulations, hexagonal prism periodic boundary conditions were applied and non bonded energies were calculated using particle mesh Ewald full electrostatics [32] (grid spacing < 0.14 nm) and a smooth (1.0–1.2 nm) cutoff of the van der Waals energy. Each system underwent 2 ps steps of energy minimization, 2 ps of equilibration at fixed volume. During this process the temperature increased from 0 to 300 K by rescaling of velocities, and the temperature was kept at 300 K by applying Langevin forces [33] to all atoms of the ZnO. The equilibration was performed using Nose´-Hoover Langevin piston pressure control at 1.0 atm for 0.5ns [34] with integration time step chosen was 1 fs.

All simulations, for both ZnO nanotube sizes, were performed at a fixed volume with the temperature maintained by Langevin dynamics applied only to the atoms of the ZnO. When a uniform electrical field is applied to all atoms, it induces, at the beginning of the simulation, a rearrangement of the ions and water focuses the electrical field to the vicinity of the ZnO nanotube. This has led to neutralizing the field in the bulk. The gradient of this electrostatic potential drives the ions current through the ZnO nanotube. A constant electric field was applied to produce the desired drifting voltage across the system. The bias voltage was applied along 11.1 nm system length in z-direction (solution - ZnO tube – solution). The present MD simulation allows a real time observation of the conformation and to determine the corresponding ionic current. The ionic current can be computed from the MD trajectory by summing up local displacements of all ions over a time interval between trajectory time frames interval Δt [35] as:

$$I\left(t + \frac{\Delta t}{2}\right) = \frac{1}{L_z \Delta t} \sum_{i=1}^{N} q_i \left(z_i(t + \Delta t) - z_i(t)\right) \qquad (39)$$

where z_i and q_i are the z coordinate and charge of ion i, respectively, N is the total number of ions, and L_z is the system simulated length along the z axis in the direction of the applied field. The interval

$z_i(t+\Delta t) - z_i(t)$ was computed respecting the periodic boundary conditions.

3.4. WATER DENSITY PROFILES

From the MD simulation trajectories, water isochore profiles are obtained by introducing a cylindrical binning, which uses the topmost ZnO layer as zero reference level and the surface normal through the center of mass of the droplet as reference axis. The density profile was computed using horizontal layers of 1 Å heights. Each layer (bin) was displaced by 0.5 Å in the Z-direction (with equal volume), i.e., the radial bin boundaries are located at $r_i = \sqrt{i \, \partial A / \pi}$ for $i=1, ..., N_{bin}$ with a base area per bin of $\partial A = 95$ Å^2, starting from the base and ending at the top of the droplet. Within each layer, the radial density was calculated using circular bins of 1 Å thickness that were concentric around the center of mass of the droplet; for each bin, the inner boundary was displaced by 0.5 Å along the horizontal axis until 20 bins with zero density were reached. To extract the water contact angle from such a profile, a two-step procedure is adopted, as described by de Ruijter et al. [36]. First, the location of the equimolar dividing surface is determined within every single horizontal layer of the binned drop. Second, a circular best fit through these points is extrapolated to the ZnO surface where the contact angle θ is measured. Note that the points of the equimolar surface below a height of 5 Å from the ZnO surface are not taken into account for the fit, to avoid the influence from density fluctuations at the liquid-solid interface. Furthermore, only those points are used for which the density measured in the central bin lies within a range of 0.5-1.1 g cm^{-3}. This effectively excludes the points in the cap region where statistics are poor. The density profile from the bulk water region across the liquid-vapor interface (z > 5 Å) fits the hyperbolic tangent functional form which represents the decrease of the radial density:

$$\alpha(z) = \frac{\alpha_1}{2}\left(1 - \tanh\left(\frac{2(z-z_e)}{w}\right)\right) \qquad (40)$$

where the vapor density is assumed to be zero, $α_l$ is the density of the bulk liquid, z_e the height (layer boundary), and w a measure for the width of the liquid-vapor interface (thickness of the interface). The layer boundary z_e resulting from matching Eq. 40 to the radial density was fitted to a circular segment, the boundary within 5 Å from the bottom being excluded.

Chapter 4

RESULTS AND DISCUSSION

4.1. DENSITY PROFILES AND WCA

After the ZnO polar surfaces were obtained and the size of the ZnO-water system was specified, we proceeded to investigate how the partial charge and force field parameters affect the wettability of different ZnO surfaces. Here, three series of MD simulations were considered within each of which the ZnO-water interaction parameters are identical, for different ZnO geometries. In the first series, (0001) and (000$\bar{1}$) ZnO polar surface slabs were used. For the second series, ZnO nanorods array were used. Finally, in the third series, ZnO nanotubes array were used. In all the wetting and electrowetting simulations we used interaction potential with $D_{ZnO} = 0.3582 \; kJ\,mol^{-1}$ and $r_{ZnO} = 2.149$ Å and similar droplets size with a uniform density of 0.997 g cm^{-3}. According to the modified Young's Eq. 3, the microscopic contact angle θ_w deviates from the macroscopic angle θ_Y due to the line tension τ. The effect of a positive line tension is to contract the droplet base and to increase the contact angle whereas a negative τ enhances wetting. However, as τ is believed to be on the order of $10^{-10} \; J/m$, the line tension is only expected to be significant for droplets with diameters below 10 *nm*. The requirements for a precise experimental determination of τ include an accurate contact angle measuring technique and the use of a highly purified liquid on an atomically smooth surface. These requirements are perfectly matched in MD simulations. A straightforward method to determine the effect of the line tension τ in MD simulations is to measure the contact angle θ_w and the contact line curvature $1/r_B$ for droplets of different sizes. Figure 14 shows the time dependence during a 2 *ns* MD simulation of the WCA for a spherical droplet of 1428 water atoms resting on ZnO hexagonal polar slab surface. Similar time dependence was seen for all ZnO polar surfaces, showing an initial decrease within the first 1 ns, followed by an equilibrium state over the next 1 ns. The WCA determined for the polar surfaces of ZnO slabs, it is seen to depend on the magnitude of the vdW interaction and is sensitive to the partial charges of zinc and oxygen atoms. In this case, the WCA indeed is seen to depend to the surface charge polarity magnitude where in case of O-polar surface the WCA is greater than 90° (hydrophobic) and in case of Zn-polar surface is less than 90° (hydrophilic) [37].

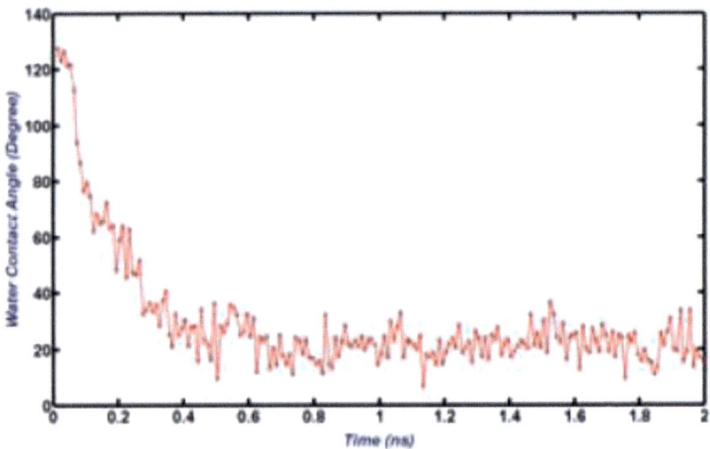

Figure 14. Variation of the contact angle during an equilibration of a water droplet on (0001)-ZnO polar slab where the initial water droplet shape is sphere with diameter of 32 Å placed on the top. The WCA values during the last 0.2ns were averaged and used as an equilibrium value.

The results described are consistent with explanations of partial charge at the surface, like dangling Zn atoms and dangling O atoms, act as hydrophilic and hydrophobic centers, respectively. Fully coordinated atoms are partially buried at the surface; hence, their electrostatic contribution is effectively screened by their neighbors. Conversely, dangling atoms are more exposed and the charges of q_O and q_{Zn} produce oriented dipoles that act as adsorption sites for water. The calculation of the water contact angle is schematically presented in figure 15.

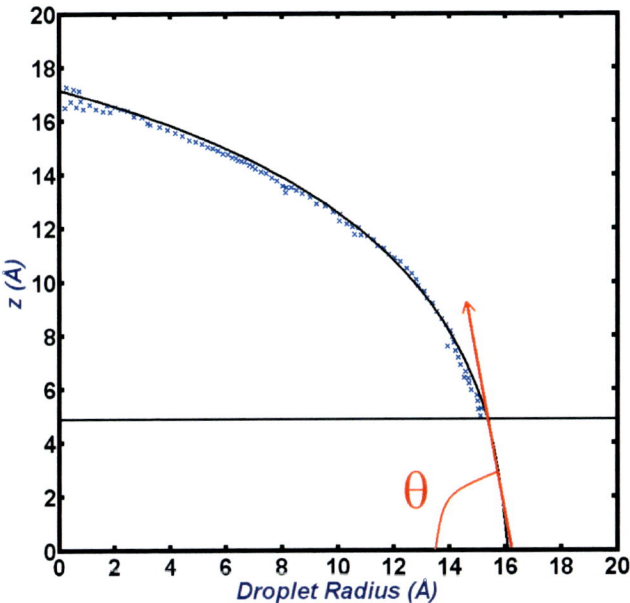

Figure 15. Water contact angle on (0001)-ZnO polar slab measured by fitting a circle with center (0,z) and radius r to the points of the equimolar planes with z>zo=5 Å from the ZnO surface is excluded from the calculations.

The density profile was computed using fitting plot (were done using the fit command and interpolation provided by MatLab v. 7.8). The WCA was averaged over the last 0.2 ns for each MD simulation (see figure 14). A few water molecules eventually evaporated from the droplet; water molecules 5 Å away from the droplet were not taken into account. It was predicted that these parameters lead to a microscopic contact angle of 20° for a 1428 water atoms droplet (TIP3P water) on (0001)-ZnO polar slab. The simulation results in a contact angle of 22.6°, which is in good agreement with the expected value [38,39]. Note that the layered structure of the liquid close to the wall (0 < z < 5 Å) is neglected in the contact angle measurement. A density profile along the centerline of the droplet is shown in figure 16.

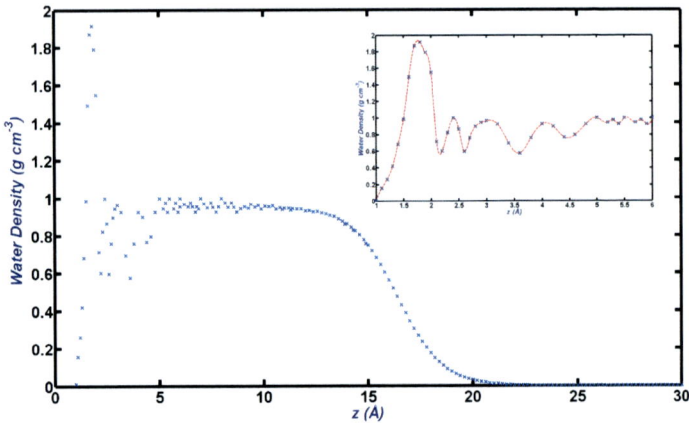

Figure 16. Water density profile for (0001)-ZnO polar slab along the centerline of the droplet; for the calculation of the contact angle only the data points contained within the boundaries (z>5 Å : 0.5<α_1(z)<1.1 g cm^{-3}).

Close to the ZnO, one pronounced density peak can be identified at distance of 1.8 Å with peak height of 1.9 $g.cm^{-3}$. A fit of Eq. 40 to the profile obtained for (0001)-ZnO polar slab yields a bulk liquid density of α_1 = 0.989 $g.cm^{-3}$ and w = 4.5 Å, resulting in an interface thickness of 4.9 Å (the interface thickness is here defined as the region where the water density drops from 0.9 α_1 to 0.1 α_1).

For the electrowetting case, we have varied the applied voltage on the system to find the values that reproduce the WCA. In this case, the positive partial charge of the Zn surface atoms on (0001)-ZnO polar slab were modulated according to the polarity of the applied voltage (+20 V) making the (0001)-Zn surface react as hydrophobic surface with WCA of 96.3° (see figure 17) [40]. Employing these applied potentials to the electrowetting model problem of a water droplet on ZnO polar slab results in qualitatively different behavior ranging from a strongly hydrophobic to hydrophilic interface [40,41] (see figure 9). These ZnO surface behaviors for switching for hydrophobic to hydrophilic faces were also are found in case of simulating the wetting and electrowetting behaviors for ZnO nanorods/tubes array with water box droplet, here the wetting mechanisms have two behaviors in the same time, spreading which is small compared with flat surfaces and filling which is depending on the nanorod size and the spaces between the nanorods/tubes for the organized array (result not shown) [42-48,40].

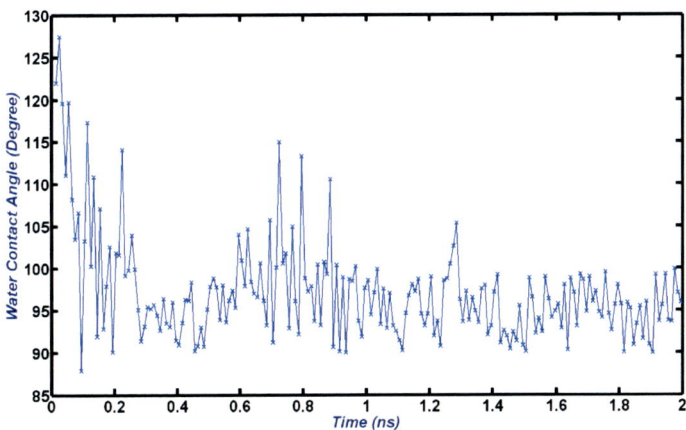

Figure 17. Variation of the contact angle during application of 20 volts on the system (water droplet on (0001)-ZnO polar slab) where the initial water droplet shape is sphere with diameter of 32 Å placed on the top. The WCA values during the last 0.2ns were averaged and used as an equilibrium value.

4.2. WATER PERMEATION THROUGH ZNO NANOTUBE

The water density in nanotube oscillates between liquid and vapour on a nanosecond time scale, a manifestation of capillary evaporation and condensation at the nanoscale. For hydrophobic tube a strong dependence of the tube state on the radius is apparent. The stable thermodynamic state switches from vapor to liquid at a critical radius $r_c = -L\cos\theta_Y$ (using $\Delta U(r_c) = 0$ in Eq. 22). The coefficient of the quadratic term, $\gamma_{lv} + (\frac{1}{2}) \Delta\mu \Delta b_{vl} L$, is positive for the hydrophobic tubes, consistent with the model Eq. 21, which predicts this coefficient to be independent of the tube inner wall. For the polar tube show that a high density of local charges leads to a higher probability of the tube being liquid filled than predicted by the macroscopic model.

For the system parameters, the coefficient is in fact dominated by the water liquid vapor surface tension γ_{lv}. $\Delta\gamma_s$, the difference in surface tensions between the tube inner wall and vapor or liquid, becomes more positive with increasing polarity of the tube wall. It effectively measures the hydrophobicity of the tube inner wall. This becomes even more apparent when the (macroscopic) contact angle $\cos\theta_Y = \Delta\gamma_s/\gamma_{lv}$ is formally computed. Macroscopically, a hydrophobic surface can be defined as one with $\theta_Y > 90^o$. A "hydrophilic" tube system is characterized by $\Delta\gamma_s > 0$ or $\theta_Y < 90^o$ and liquid is always the preferred phase in the inner tube, regardless of inner tube radius [18].

Further tests of Eq. 22 by varying the length of the tube together with the radius confirm the model qualitatively (data not shown). Our model implies that for nanoscale tubes the cost of creating the liquid-vapor interface is the only force driving the filling of a hydrophobic ($\Delta\gamma_s \leq 0$) tube. Little free energy $\Delta\mu\,\Delta b_{vl}\,\pi\,r^2 L$ is gained by creating a bulk-like liquid in the tube instead of vapor.

In the electro-simulations we control the applied voltage; the system was simulated for 1 ns. In the forward biasing first case, water completely permeated the nanotube; in the backward biasing second case, water did not penetrate the tube completely. We conclude that ZnO nanotube wall act as a capacitor inside the nanotube which can be responsible for the initial water permeability of the tube; rather the surface hydrophobicity slows down the wetting. To further test the effect of the surface hydrophobicity on wetting, we increased the hydrophobicity of the ZnO nanotube by increasing the applied voltage and analyzed the permeation kinetics. In this case, we observe that *Zn* and O surface atoms affect the permeation speed. Furthermore, the velocity of the permeation is faster for tubes with a larger diameter. In these two cases, the permeation behavior can be explained in terms of cohesive (water-water) and adhesive (water-ZnO) forces [19].

4.3. SALT CONCENTRATION DEPENDENCE ON ZnO NANOTUBE IONIC CURRENTS

In principle, three mechanisms contribute to electrolyte transport through the nanotube, these are: electrophoretic ion migration, convection (i.e., electro-osmosis), and diffusion. However, the contributions of convection and diffusion are marginal, and the contribution from ion migration predominates. As a first step toward understanding the modulation of ionic conductivity, we simulate ion transport in hexagonal ZnO nanotube with cylindrical inner shape having radius 1.46 *nm* by using MD simulations. We further calculate the ionic current as a function of the applied biasing voltage for constant electrolyte concentrations and the ionic conductance as a function of the electrolyte concentration for a constant tube inner diameter. The intention is to infer the conductivity of the electrolyte solution for the case of electrolyte solution flow in ZnO nanotube with fixed surface charges.

The dc electrolytic current through a single ZnO nanotube is calculated as a function of the applied electrochemical potentials for 20, 15, 10, 5, 4, 3, 2, 1, -1, -2, -3, -4, -5, -10, -15, -20 volt in the z-direction at 27 °C. Starting from a random configuration, the systems were simulated for 0.5 ns to reach a steady state, followed by a run of 2 ns. The ion current was computed using Eq. 39. The current–voltage (I–V) characteristics obtained from ZnO tube with inner radius of 1.46 nm, calculated over a range of electrolyte concentrations 0.5, 1.0, 5.0, 10.0 M. Notice that in all cases, the I–V characteristics are approximately linear which means that they have ohmic behavior (see figure 18a, b). By fitting a line to the data in figure 18 and taking the slope we can obtain the conductance (see figure 19), which is function of ion concentrations for a constant ZnO nanotube diameter. However, it is apparent that the conductance increased exponentially with increasing the ion concentration to 5M, after this concentration value the conductance appear to have stable behavior. It is appear from the figure 19 that the highest ionic conductance is for KCl solution then NaCl, $CaCl_2$ and $MgCl_2$ which are in a good agreement with the ionic size for these ions [49].

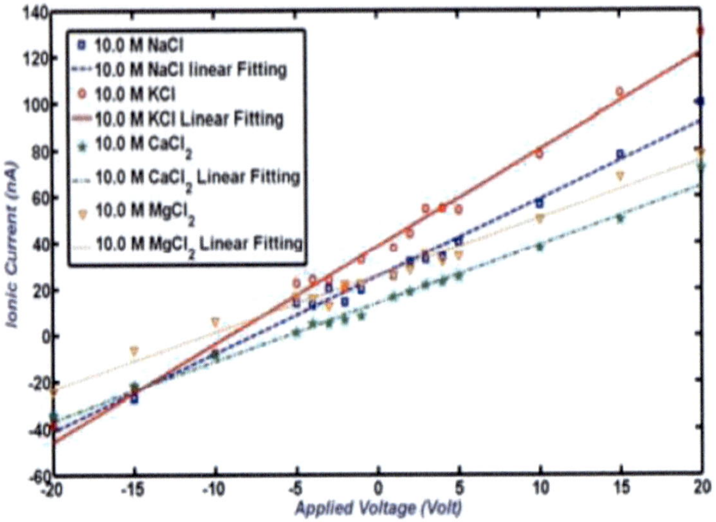

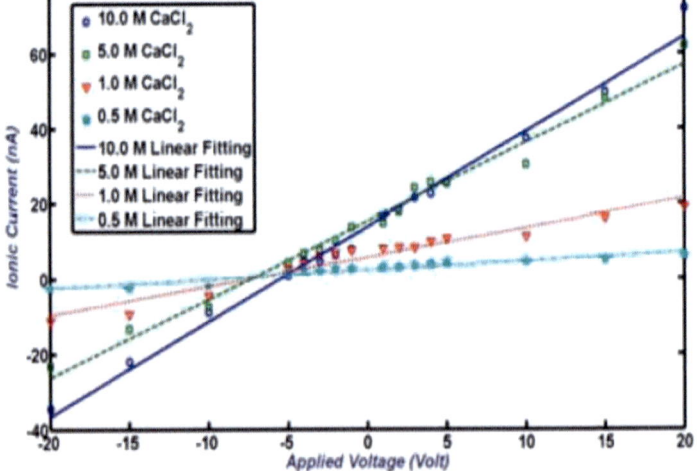

Figure 18. Ionic current as a function of applied bias voltage for (a) 10M electrolyte concentration of NaCl, KCl, $CaCl_2$, and $MgCl_2$ and for 0.5, 1.0, 5.0, and 10.0 M $CaCl_2$ (b).

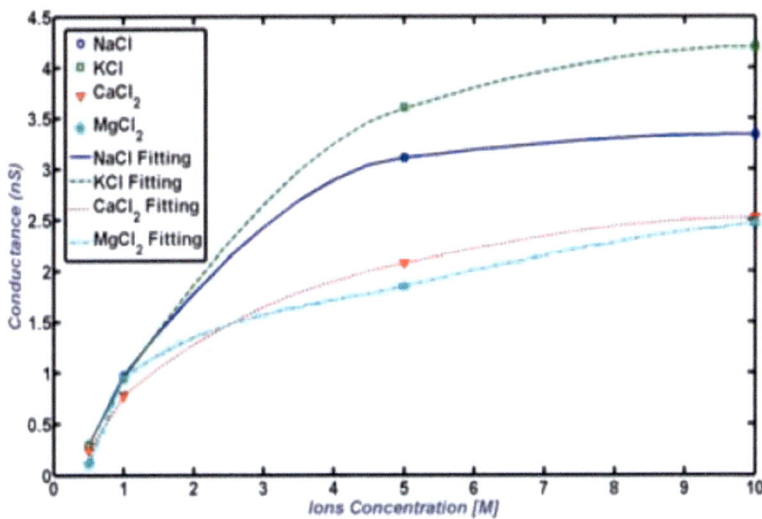

Figure 19. Ionic conductance as a function of ion concentration for NaCl, KCl, CaCl$_2$, and MgCl$_2$ electrolyte solutions.

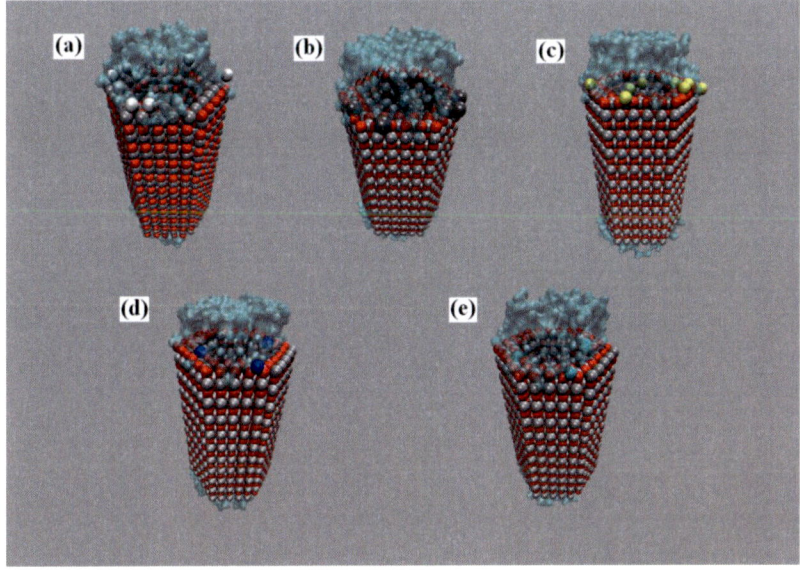

Figure 20. Adsorptions of Cl ions on the Zn-polar surface (a) and adsorption of Na, K, Ca, and Mg ions on the O-polar surface (b), (c), (d) and (e), during the MD simulations run for 5M electrolyte solution.

In the MD simulations we observed adhesion of flowing ions to the top, the bottom and to the inner surfaces of the ZnO nanotube. This adhesion will lead to reduce the flowing ionic current value. The observed adhesion was dominated by hydrophobic attraction of the nanotube walls with the water molecules and ions creating the interface region and the double layer capacitance. Figure 20 shows snap shot images of the ions movement in the ZnO nanotube (r_{tube}=1.46 nm) obtained from the MD simulations, indicating that the positions and the amount of ions for 5M (NaCl, KCl, CaCl$_2$, and MgCl$_2$) are adsorbed on the $(000\bar{1})$-O and (0001)-Zn permanent polar surfaces, respectively. Both of them were adsorbed on the dangling bonds of the nanotube inner wall surfaces depending on the electrolyte concentrations and the applied voltage. It is important to note that in our case we have a nanotube with a very thin thickness so that the permanent polar surface charge value can be reduced to include approximately the top and bottom edges of the inner tube surface. It is known that ZnO nanotubes surfaces carry a small net charge [50,51], the partial charge on the surface atoms can generate localized strong electric field that attracts ions.

To understand the scaling behavior of the ionic conductance, changes in the electrolyte concentration and ionic mobility have to be taken into account. To calculate the mobility of cation μ_{c^+} and anion μ_{a^-} we used the following equations [21,52]:

$$N_{c^+} = \frac{1}{e\mu_{c^+}\rho} \qquad (41)$$

$$N_{a^+} = \frac{1}{e\mu_{a^+}\rho} \qquad (42)$$

where N_{c^+} and N_{a^-} are the number of cation and anion, and ρ is the resistivity of electrolyte solution channel under zero gate voltage is determined from [21]:

$$\rho = \frac{dV_{DS}}{dI_{DS}} \frac{A}{L} \qquad (43)$$

where A is the channel cross-section area and L is the length of the channel between the source and drain electrodes.

The adsorption of cation and anion at the inner wall surface over screens the surface charge. Notice that the cation and anion concentration are oscillates with distance from the inner wall of the tube. The ion mobility gradually increases as it moves away from the inner tube surface toward the center of the tube. In the larger tube inner diameter (D_{in} = 3.3 nm), the mobilities near the center of the tube are almost the same and are comparable with the bulk value, whereas for smaller inner diameter (D_{in} = 2.0 nm) tube, the ions mobility in the tube center is smaller of its bulk value (see figure 21) [53]. The reduction in the ion mobilities are observed, indicating that the ion mobility in a nanotube with inner diameter D_{in} = 2.0 nm is substantially affected by confinement and ion–surface interactions [53].

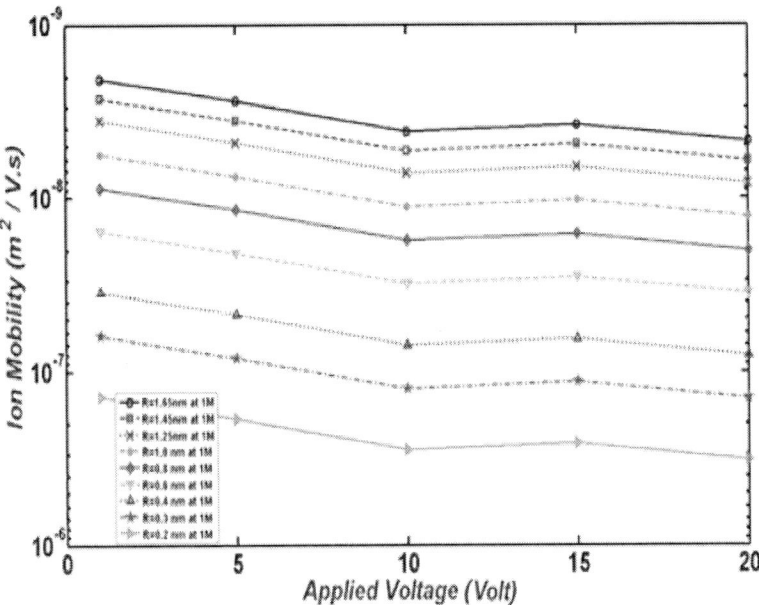

Figure 21. Ion mobility as a function of the applied bias voltage for ZnO nanotube R_{in}=1.65nm at 1M KCl with changing the distance from the tube inner surface [53].

Chapter 5

CONCLUSION

We have explicitly demonstrated a systematic study of the potential functions used to model the interaction between water and ZnO nanostructures in molecular dynamics simulations.

Our results were performed to analyze the behavior of water and ions adsorbed on ZnO surfaces on the nanosecond scale. Three different geometries were considered in the study: (0001)-Zn and (000$\bar{1}$)-O polar surface slabs, nanorod crystals, and nanotubes were modeled. We have used UFF force field parameters and refined them to fit the CHARMM force field parameters for ZnO surfaces that simulate wetting properties with different ZnO nanostructure films based on the observation of WCA. The water density profiles were analyzed and yielded important results. A hydrophobic mechanism was found for water droplet interaction with (000$\bar{1}$)-ZnO polar slab with contact angle 96° (water moves away from the surface) and a hydrophilic mechanism for the (0001)-ZnO polar slab with contact angle 54° (overlap between water and the surface atoms) for the equilibrium case as results of partial surface charges interactions acting between the oxygen atoms of the water and Zn atoms sites for (0001) surface or O atoms sites for (000$\bar{1}$) surface. An applied voltage on the system were used to simulate the electrowetting behavior, these interactions shows, for the case of (0001)-ZnO polar slab when we applied backward biasing voltage, the system still have hydrophilic face with WCA less than 20°, on the other hand when we forward biasing voltage the system will have hydrophobic face with WCA of 96.3°. These ZnO surface behaviors for switching from hydrophobic to hydrophilic faces were also founded in cases of simulated the wetting and electrowetting behaviors for ZnO nanorods/tubes array where the wetting mechanisms have two behaviors in the same time, spreading and filling.

The results from the permeation of water through ZnO nanotube model are an important factor to explain the initial hydrophobicity and hydrophilicity of the ZnO nanotubes for equilibrium and applied voltage cases. A thermodynamic model based on surface energies fits remarkably well with the data of the atomic-scale MD simulations. Our simulations show that the polarity of the tube wall can shift the WCA considerably. By applying voltage across the system the hydrophobic face are rotated to hydrophilic face by combining hydrophobic face with a change in surface polarity. It appears that in this case water structure (i.e. a hydrogen bond network) is responsible for different wall-fluid surface tensions. As a result, a moderate change in applied voltage is required to obtain the required face behavior for different applications.

We have inferred the conductance and ionic mobility of NaCl, KCl, $CaCl_2$, and $MgCl_2$ electrolytes flowing through ZnO nanotubes. We performed calculations of the ionic current as a function of the electrolytes concentration. The nanotube conductance variation versus the concentrations are found to change nonlinearly with increasing the electrolyte concentration and after 5M the conductance have reached saturation values for the nanotube diameter (3.2 nm). We interpret these observations by using MD simulations of the ion transport through the ZnO nanotubes and found that the calculated conductance are consistent with the presence of fixed surface charges in the nanotube inner surface. Moreover, a reduction of the ion mobility due to this surface charge is also observed. We conclude that the ionic current in ZnO nanotube are governed by fixed surface charge layer and can be modulated by applying voltage across the ZnO nanotube outer walls (V_{gate}). In this case a ZnO nanotube can be operated in a way analogous to FET operation leading to control of ionic fluid flow through nanotubes [54-56]. Such control of ionic flow would be of interest to many biological systems. This model will be used to simulate the ion selective conductance through ZnO nanotubes and to calculate the binding infinities of various biological macromolecules to ZnO surfaces.

REFERENCES

[1] Kumar, S. A.; Chen, S.-M. *Analyt. Lett.* 2008, *41*, 141-158.
[2] Wei, A.; Sun, X.W.; Wang, J.X. *Appl. Phys. Lett.* 2006, *89*, 123902.
[3] Yeh, P.-H.; Li, Z.; Wang, Z. L. *Adv. Mater.* 2009, *21*, 1–4.
[4] Yang, K.; She, G.-W.; Wang, H.; Ou, X.-M.; Zhang, X.-H.; Lee, C.-S.; Lee, S.-T. *J. Phys. Chem. C*, 2009, DOI: 10.1021/jp901894j.
[5] Al-Hilli, S.; Willander, M. *Nanotechnology* 2009, *20*, 175103.
[6] Al-Hilli, S. M.; Öst, A.; Strålfors, P.; Willander, M. *J. Appl. Phys.* 2007, *102*, 084304.
[7] Al-Hilli, S.; Willander, M. *Sensors* 2009, *9*, 7445-7480.
[8] Young, T. *Philos. Trans. R. Soc. London* 1805, *95*, 65-87.
[9] Wang, J. Y.; Betelu, S.; Law, B.M. *Phys. Rev. E* 2001, *63*, 031601.
[10] Adamson, A. W. *Physical Chemistry of Surfaces*; 6th edit; John Wiley & Sons: New York, 1997, pp 195-197.
[11] Welters, W. J. J.; Fokkink, L. G. J. *Langmuir* 1998, *14*, 1535-1538.
[12] Mugele, F.; Baret, J.-C. *J. Phys.: Condens. Matter* 2005, *17*, R705–R774.
[13] Wenzel, R. N. *Ind. Eng. Chem.* 1936, *28*, 988–994.
[14] Cassie, A. B. D.; Baxter, S. *Trans. Faraday Soc.* 1944, *40*, 546 – 551.
[15] Han, J.; Gao, W. *J. Electron. Mater.* 2009, *38*, 601-608.
[16] Beckstein, O.; Sansom M. S. P. *Phys. Biol.* 2004, *1*, 42-52.
[17] de Gennes, P. G. *Rev. Mod. Phys.* 1985, *57*, 827–863.
[18] Allen, R.; Hansen, J. P.; Melchionna S. *J. Chem. Phys.* 2003, *119*, 3905–3919.

[19] Heikenfeld, J.; Zhou, K.; Kreit, E.; Raj, B; Yang, S.; Sun, B.; Milarcik, A.; Clapp, L.; Schwartz, R. *Nature Photonics* 2009, *3*, 292-296.
[20] Vidal, J.; Gracheva, M. E.; Leburton, J.-P. *Nanoscale Res. Lett.* 2007, *2*, 61–68.
[21] Sze, S. M. *Physics of Semiconductor Devices*; Wiley: New York, 1981.
[22] Phillips, J. C.; Braun, R.; Wang, W.; Gumbart, J.; Tajkhorshid, E.; Villa, E.; Chipot, C.; Skeel, R. D.; Kale, L.; Schulten, K. *J. Comput. Chem.* 2005, *26*, 1781-1802 http://www.ks.uiuc.edu/Research/namd/.
[23] Matlab, v. 7.5; The MathWorks, Inc., 2007.
[24] Humphrey, W.; Dalke, A.; Schulten, K. *J. Molec. Graphics* 1996, *14*, 33-38 http://www.ks.uiuc.edu/Research/vmd/.
[25] Rappé, A. K.; Casewit, C. J.; Colwell, K. S.; Goddard, W. A. I.; Skiff, W. M. 1992 *J. Am. Chem. Soc.* 1992, *114*, 10024-10035.
[26] MacKerell, A. D.; Brooks, B.; Brooks III, C. L.; Nilsson, L.; Roux, B.; Won, Y.; Karplus, M. *CHARMM: The energy function and its parameterization with an overview of the program*. In *The Encyclopedia of Computational Chemistry*; P. Schleyer; Ed.; John Wiley & Sons: UK, 1998, pp. 271–277.
[27] Schulz, H.; Thiemann, K. H. *Solid State Commun.*1979, *32*, 783.
[28] Werder, T.; Walther, J. H.; Jaffe, R. L.; Halicioglu, T.; Koumoutsakos, P. *J. Phys. Chem. B* 2003, *107*, 1345-1352.
[29] Cruz-Chu, E. R.; Aksimentiev, A.; Schulten, K. *J. Phys. Chem. B* 2006, *110*, 21497-21508.
[30] Martyna, G. J.; Tobias, D. J.; Klein, M. L. *J. Chem. Phys.* 1994, *101*, 4177–4189.
[31] Shimomura, K.; Nishiyama, K.; Kadono, R. *Phys. Rev. Lett.* 2002, *89*, 255505.
[32] Batcho, P. F.; Case, D. A.; Schlick, T. *J. Chem. Phys.* 2001, *115*, 4003–4018.
[33] Brunger, A. T. *X-PLOR*, Version 3.1: *A System for X-ray Crystallography and NMR*; The Howard Hughes Medical Institute and Department of Molecular Biophysics and Biochemistry, Yale University, 1992.
[34] Martyna, G. J.; Tobias, D. J.; Klein, M. L. *J. Chem. Phys.* 1994, *101*, 4177–4189.

References

[35] Aksimentiev, A.; Schulten, K. *Biophys. J.* 2005, *88*, 3745–3761.
[36] de Ruijter, M. J.; Blake, T. D.; De Coninck, J. *Langmuir* 1999, *15*, 7836-7847.
[37] Wu, P. C.; Losurdo, M.; Kim, T.-H.; Giangregorio, M.; Bruno, G.; Everitt, H. O.; Brown, A. S. *Langmuir* 2009, *25*, 924-930.
[38] Zhang, Z.; Chen, H.; Zhong, J.; Saraf, G.; Lu, Y. *J. Electron. Mater.* 2007, *36*, 895-899.
[39] Luo, J. *Langmuir* 2005, *21*, 7358-7365.
[40] Campbell, J. L.; Breedon, M.; Latham, K.; Kalantar-zadeh K. *Langmuir* 2008, *24*, 5091-5098.
[41] Shamai, R.; Andelman, D.; Bergec, B.; Hayes, R. *Soft Matter* 2008, *4*, 38–45.
[42] Li, G.; Chen, T.; Yan, B.; Ma, Y.; Zhang, Z.; Yu, T.; Shen, Z.; Chen, H.; Wu, T. *Appl. Phys. Lett.* 2008, *92*, 173104.
[43] Zhang, J.; Huang, W.; Han, Y. *Langmuir* 2006, *22*, 2946-2950.
[44] Sun, M.; Du, Y.; Hao, W.; Xu, H.; Yu, Y.; Wang, T. *J. Mater. Sci. Technol.* 2009, *25*, 53-57.
[45] Wu, X.; Zheng, L.; Wu, D. *Langmuir* 2005, *21*, 2665-2667.
[46] Zhang, X.-T.; Sato, O.; Fujishima, A. *Langmuir* 2004, *20*, 6065-6067.
[47] Feng, X.; Roy, S. C.; Grimes, C. A. *Langmuir* 2008, *24*, 3918-3921.
[48] Feng, X.; Feng, L.; Jin, M.; Zhai, J.; Jiang, L.; Zhu, D. *J. Am. Chem. Soc.* 2004, *126*, 62-63.
[49] Zumdahl, S. S. *Chemical Principles*; 5th Edit.; Houghton Mifflin Company: USA, MA, 2005.
[50] Wander, A.; Harrison, N. M. *J. Chem. Phys.* 2001, *155*, 2312-2316.
[51] Kresse, G.; Dulub, O.; Diebold, U. *Phys. Rev. B* 2003, *68*, 245409.
[52] Muller, R. S.; Kamins, T. I.; Chan, M. *Device Electronics for Integrated Circuits*; Wiley: New York, 2003.
[53] Al-Hilli, S.; Willander, M. *Nanotechnology* 2009, *20*, 505504.
[54] Karnik, R.; Fan, R.; Yue, M.; Li, D.; Yang, P.; Majumdar, A. *Nano Lett.* 2005, *5*, 943-948.
[55] Vermesh, U.; Choi, J. W.; Vermesh, O.; Fan, R.; Nagarah, J.; Heath, J. R. *Nano Lett.* 2009, *9*, 1315-1319.
[56] Lee, C.-S.; Kim, S. K.; Kim, M. *Sensors* 2009, *9*, 7111-7131.

INDEX

A

adhesion, 37
adsorption, 8, 33, 39, 40
applications, vii, 2, 3, 6, 9, 17, 44
atoms, 19, 20, 21, 22, 23, 24, 27, 28, 32, 33, 34, 36, 38, 43

B

behavior, vii, 9, 13, 25, 34, 36, 37, 40, 43, 44
bending, 18, 19, 20
bias, 24, 28, 38, 41
binding, 1, 44
biochemistry, 1
biological systems, 44
biosensors, 3
Boltzmann distribution, 7
bonds, 19, 20, 22, 27, 37

C

calcium, 27
capillary, 35
case study, 1
cation, 15, 40
cell, 1, 21
cell metabolism, 1
channels, 2, 3
character, 11, 26
charge density, 7
chemical stability, 21
complement, 17
concentration, 15, 17, 27, 37, 38, 39, 40, 44
condensation, 35
conductance, vii, 16, 37, 39, 40, 44
conductivity, 37
configuration, 23
confinement, 13, 40
conjugate gradient method, 23, 24
consumption, 3
control, 3, 28, 36, 44
crystals, 43

D

deficiencies, 1
definition, 13
density, 11, 15, 17, 22, 28, 29, 31, 33, 34, 35, 43
density fluctuations, 29
detection, 1
dielectric constant, 7, 19
diffusion, 15, 37
distribution, 6, 7

divergence, 15
dyes, 1
dynamics, vii, 14, 17, 28

E

electric field, 3, 6, 8, 13, 15, 24, 28, 38
electrodes, 15, 40
electrolyte, vii, 7, 14, 27, 36, 37, 38, 39, 40, 44
energy, vii, 4, 7, 12, 18, 19, 20, 23, 27, 46
environment, 1, 4, 17
equilibrium, vii, 4, 5, 13, 19, 32, 35, 43, 44
evaporation, 35

F

films, 9, 17, 43
fluid, 44
force constants, 19
free energy, 4, 12, 13, 36
freedom, 19

G

genetics, 1
graph, 13

H

height, 10, 26, 29, 34
hybrid, 23
hydrogen, 44
hydrophobicity, 3, 10, 36, 44

I

images, 37
indices, 4
insertion, 1

integration, 23, 28
interactions, vii, 1, 4, 17, 19, 20, 23, 40, 43
interface, 3, 6, 7, 8, 9, 10, 13, 29, 34, 36, 37
intermolecular interactions, vii
interval, 28
ion transport, 37, 44
ionicity, 20
ions, vii, 1, 6, 7, 11, 15, 22, 27, 28, 37, 39, 40, 43

K

kinetics, vii, 36

L

lattice parameters, 21
line, 4, 6, 17, 31, 37
liquid-air interface, 9
liquids, 6
Luo, 47

M

macromolecules, 44
magnesium, 27
measurement, 33
measures, 36
methodology, 1
migration, 36
mobility, 14, 15, 40, 41, 44
model, 7, 9, 11, 12, 13, 14, 20, 22, 26, 34, 35, 36, 43, 44
models, 17
molecular dynamics, vii, 1, 14, 18, 43
molecules, 1, 3, 6, 11, 12, 22, 24, 26, 27, 33, 37
morphology, 17, 20
motion, 22
movement, 3, 37

Index

N

NaCl, vii, 17, 27, 37, 38, 39, 44
nanofabrication, 1
nanometer, 8, 10
nanorods, vii, 10, 11, 22, 24, 35
nanostructures, 9
nanotechnology, vii
nanotube, vii, 1, 11, 12, 14, 15, 16, 17, 18, 21, 22, 26, 27, 28, 35, 36, 37, 40, 41, 44
neglect, 8
network, 44
NMR, 46

O

observations, 44
one dimension, 15
openness, 13
order, 6, 7, 26, 31
osmosis, 37
overlap, 43
oxygen, 20, 22, 32, 43

P

parallel, 26
parameters, vii, 19, 20, 26, 31, 33, 36, 43
pathways, 1
performance, 3
permeation, vii, 11, 14, 17, 26, 36, 44
permittivity, 20, 27
physics, 2
polarity, 6, 20, 32, 34, 36, 44
polarization, 20
poor, 29
potassium, 27
pressure, 28
probability, 35
probe, 1
program, 18, 21, 46
properties, 1, 3, 4, 9, 13, 18, 20, 43

R

radius, 4, 10, 11, 13, 17, 20, 21, 22, 26, 33, 35, 36, 37
random configuration, 37
range, vii, 23, 29, 37
reagents, 1
real time, 28
recombination, 15
recommendations, iv
region, 6, 29, 34, 37
resistance, 15
resolution, 1
rods, 1, 9, 10, 11, 17, 21, 22
roughness, 9, 11
routines, 18

S

salt, 6, 17
saturation, 12, 44
scaling, 40
semiconductor, 8, 19, 20
sensing, 1, 3, 9
sensitivity, 1, 3, 17
sensors, vii, 1
separation, 22
shape, 4, 11, 17, 21, 22, 26, 32, 35, 37
simulation, 11, 17, 18, 24, 27, 28, 32, 33
sodium, 27
solid phase, 3
solid surfaces, 4, 10
speed, 36
stability, 21
standard deviation, 23
statistics, 29
stretching, 18, 19
substrates, 6
Sun, 45, 47
surface structure, 3
surface tension, 4, 36, 44
switching, 23, 34, 43
symmetry, 20

T

temperature, 23, 28
tension, 4, 6, 17, 31
tensions, 3, 36
thermodynamics, 6
time frame, 28
trajectory, 13, 23, 24, 28
transistor, vii
transition, 3, 27
transition rate, 3
transport, 1, 36

U

uniform, 24, 28, 31

V

vacuum, 20

vapor, 4, 11, 12, 29, 35, 36
variations, 15
velocity, 36

W

water permeability, 36
wettability, 3, 5, 6, 9, 17, 20, 31
wetting, 1, 3, 4, 5, 6, 9, 13, 17, 18, 22, 25, 31, 34, 36, 43

Z

zinc, 20, 22, 32
ZnO, i, iii, vii, 1, 3, 9, 10, 11, 12, 14, 17, 19, 20, 21, 22, 23, 24, 25, 26, 27, 28, 31, 32, 33, 34, 35, 36, 37, 41, 43, 44
ZnO nanorods, 10, 11, 21, 22, 23, 24, 25, 31, 34, 44
ZnO nanostructures, 1, 3, 17, 24, 43